GROWING THE CITRUS FRUITS

With Information on Growing Lemons, Oranges, Grape Fruits and Other Citrus Fruits

BY

GEORGE W. HOOD

British Library Cataloguing-in-Publication Data
A catalogue record for this book is available from the
British Library

CONTENTS

Introduction to Fruit Growing

In botany, a fruit is a part of a flowering plant that derives from specific tissues of the flower, one or more ovaries, and in some cases accessory tissues. In common language use though, 'fruit' normally means the fleshy seed-associated structures of a plant that are sweet or sour, and edible in the raw state, such as apples, oranges, grapes, strawberries, bananas, and lemons. Many fruit bearing plants have grown alongside the movements of humans and animals in a symbiotic relationship, as a means for seed dispersal and nutrition respectively. In fact, humans and many animals have become dependent on fruits as a source of food. Fruits account for a substantial fraction of the world's agricultural output, and some (such as the apple and the pomegranate) have acquired extensive cultural and symbolic meanings. Today, most fruit is produced using traditional farming practices, in large orchards or plantations, utilising pesticides and often the employment of hundreds of workers. However, the yield of fruit from organic farming is growing – and, importantly, many individuals are starting to grow their own fruits and vegetables. This historic and incredibly important foodstuff is gradually making a come-back into the individual garden.

The scientific study and cultivation of fruits is

called 'pomology', and this branch of methodology divides fruits into groups based on plant morphology and anatomy. Some of these useful subdivisions broadly incorporate 'Pome Fruits', including apples and pears, and 'Stone Fruits' so called because of their characteristic middle, including peaches, almonds, apricots, plums and cherries. Many hundreds of fruits, including fleshy fruits like apple, peach, pear, kiwifruit, watermelon and mango are commercially valuable as human food, eaten both fresh and as jams, marmalade and other preserves, as well as in other recipes. Because fruits have been such a major part of the human diet, different cultures have developed many varying uses for fruits, which often do not revolve around eating. Many dry fruits are used as decorations or in dried flower arrangements, such as lotus, wheat, annual honesty and milkweed, whilst ornamental trees and shrubs are often cultivated for their colourful fruits (including holly, pyracantha, viburnum, skimmia, beautyberry and cotoneaster).

These widespread uses, practical as well as edible, make fruits a perfect thing to grow at home; and dependent on location and climate – they can be very low-maintenance crops. One of the most common fruits found in the British countryside (and towns for that matter) is the blackberry bush, which thrives in most soils – apart from those which are poorly drained or mostly made of dry or sandy soil. Apple trees are, of course, are another classic and whilst they may take several years to grow into a well-established tree, they will grow nicely in most sunny and well composted areas. Growing one's own fresh, juicy tomatoes is one of the great pleasures of summer gardening, and even if the gardener doesn't have room for rows of plants, pots or hanging baskets are a fantastic solution. The types, methods and approaches to growing fruit are myriad, and far too numerous to be discussed in any detail here, but there are always easy ways to get started for the complete novice. We hope that the reader is inspired by this book on fruit and fruit growing – and is encouraged to start, or continue their own cultivations. Good Luck!

THE CITRUS FRUITS

THE citrus fruits include the orange, lemon, grape fruit, tangerine, kumquat and lime. All of these fruits are grown in the citrus belt, but the oranges, lemons and the grape fruit are the most important of the citrus fruits.

The citrus industry is confined to certain definite regions of the United States. The citrus regions are located in California, Florida, Texas, New Mexico and Arizona, but by far the larger commercial plantings are found in California and in Florida. The original home of the citrus fruit was in India and the Malay Archipelago, but today the great bulk of the oranges which supply the markets of the world are produced in California, Florida, Spain, Palestine, Australia, Italy and Japan. Certain parts of Mexico produce citrus fruit to a limited extent, but they seem to lack good shipping qualities.

The citrus fruit was first introduced into this country in southern California. The seed was brought into this state from the lower peninsula of California by the early Spanish settlers. These settlers not only brought the citrus fruit but they introduced many other tropical and semitropical fruits such as figs, grapes, olives and dates.

Orange.—The orange is divided into several species, some of which are edible, as the sweet oranges, and others, the sour oranges, which are used for the manufacture of certain oils and other products.

The sweet orange is the one which is commonly known and which is used extensively for the table. The sweet orange includes many varieties. The Navel and the Valencia are the most important varieties, although there is a long list, and some give excellent promise for the production of superior fruit.

The sour orange is used principally as stock on which to graft the sweet varieties. The fruit of the sour orange is not edible, but it is used to some extent for flavoring. The sour orange is valuable as a stock because of its resistance to the foot rot and the gum diseases.

Grape Fruit.—Technically the term grape fruit is incorrect, but it has gained so much prestige on the market that it will in all probability remain. The correct name of this fruit is pomela. The grape fruit is usually a prolific bearer. The fruit is gaining rapidly in popularity and more of it is consumed every year.

Kumquat.—The kumquat is a small yellow citrus fruit resembling a small orange. It is sometimes called golden orange. The fruit is often used for decorating and the pulp of the fruit for preserving.

Lime.—The lime is a small yellow citrus fruit resembling a small lemon. It is classed as a shrub, but when it is given room to grow it forms a small tree. The lime is the most tender of the citrus fruits, and it is killed back by a slight frost but usually sprouts up vigorously the next year. The skin of the lime is thin and of a lemon-yellow color. The pulp is a pale green and is filled with a very sharp acid Juice. The juice and the pulp of the lime are better for most purposes than that of the lemon, and it is used in preference to the lemon by people in tropical countries. The lime is now found on most of our important northern markets, due to the better transportation facilities.

Lemon.—The lemon is one of the best-known citrus fruits. While the lemon is not as extensively cultivated as the orange it is probably equally as valuable. The lemon is gaining in popularity in the citrus regions and it is now being planted more than formerly. The lemon is supposed to have originated from the citron, and it was first introduced into Palestine and Egypt in the tenth century and into Europe at the time of the crusades. The lemon trees are faster growing than the orange trees, and they are usually more productive and will stand more neglect. The lemon includes both the sweet and the sour types.

Propagation.—The citrus fruits are usually propagated by budding. Although a few varieties can be grown with some success by grafting, cutting and layering these methods are not to be generally recommended. Occasionally the lemon can be grown from cuttings, but the trees are rarely ever successful. The orange cutting can rarely ever be made to take root, and this method of propagation should not be considered.

FIG. 123.—Making a new top on a citrus tree.

The shield or T-bud is universally used in the propagation of the citrus fruits. The success of this method is largely dependent upon the proper selection of the budding wood. Citrus buds should be cut from round plump wood taken from fruiting branches. Suitable wood of this character is difficult to find on the orange but it is more plentiful on the lemon. The orange tree from which the buds are to be selected must be prepared a year in advance. The preparation of the tree consists in systematically pruning the branches to a given length and causing healthy, plump buds to form.

The bud-sticks are usually all cut at one time and stored until they are used. It is thought by some nurserymen that bud-sticks which have been stored produce a larger percentage of buds that will grow. Stored bud-sticks produce a greater percentage of uniform trees because the buds seem to mature in some way during storage. The bud-sticks are tied in bundles and either packed in damp sand, sawdust or damp moss until they are used. The damp moss or sawdust is preferred by most growers, because the sand has a tendency to dull the budding knife. The bud-sticks should have the leaves pruned off, leaving a little of the leaf stem to serve as a handle with which to hold the bud.

FIG. 124.—The method of top working a citrus orchard.

The budding of the citrus fruit is usually performed during November or December in those regions where a more or less definite winter occurs. This is called dormant budding. If any buds fail to grow from dormant budding or are killed back by the winter then spring budding is performed. Spring budding should be done after a vigorous growth has started. However, budding can usually be performed at any time of the year when the bark peels or separates easily from the wood.

Soil.—The citrus fruits arc very cosmopolitan with regard to soil. In California as well as in Florida and other citrus regions the citrus fruits are grown on a great variety of soils, ranging from light sandy soil through loams to black heavy, adobe soils. There are probably no other fruit trees which are

10

so plastic and which will adapt themselves with such ease as to grow on almost any type of soil. The determining factor in the soil seems to be its physical condition and where this is good the citrus fruit is almost sure to grow on any soil. From the standpoint of cultivation and ease of handling the soft sandy loams should be preferred over the sticky, heavier soils.

The subsoil perhaps influences the growth of the citrus fruits to a greater extent than does the top soil. In many cases the layer of soil just below the top soil varies in thickness from a few inches to several feet. Sometimes this subsoil is so hard and so firm as to be impervious to water. Not only does it prevent the water from soaking away or rising from lower levels, but it is so hard that the roots of the plants cannot penetrate it. On such a soil the root development of the tree is greatly restricted.

Sometimes the subsoil is too loose and open and unretentive of moisture. This condition gives a deficiency of plant food and a lack of water.

A good soil then for the citrus tree may be of any type, but preferably a sandy loam which should be at least 4 or 5 feet deep. It must be well drained and the subsoil should neither be too hard nor very loose.

Cultivation.—Good preparation of the soil is essential if profitable crops are to be grown. The soil should be thoroughly plowed and worked into a very fine state before any trees are planted. After the trees are set, the soil should be plowed thoroughly once a year, preferably in March or April. At this time the cover crop should be turned under. The plowing should be completed before the tree comes into full bloom, in order to avoid the cutting of the roots at this critical time. The depth of plowing should vary with each year so as to prevent any hard layer from forming by the pressure of the plow.

The proper cultivation determines the success of the orchard. It makes little difference whether the trees are grown on irrigated land or on dry land, frequent cultivations should be given to the soil. The soil should be stirred to a depth of about 4 inches after each irrigation or after each rain. No attempt should be made to cultivate until after the soil has dried out, or until it is in the proper condition for cultivation. If the soil is cultivated when it is too wet it will be hard and lumpy.

FIG. 125.—Protecting a young orange tree from the hot
sun.

The tillage implement should be selected with reference to
the type of soil. One type of soil will require one kind of a
cultivator and another type of soil will need a different tool.

13

Planting.—The planting of a citrus tree is similar to that of any other fruit tree. The general conditions concerning the preparation of the soil, the digging of the holes, etc., are identical to those of other fruits.

The distance to set the trees is determined by the variety as well as by the fertility and the character of the soil. The smaller growing varieties such as the Mandarin oranges and the limes should not number more than 200 trees to the acre. This will mean that the trees should be set about 12 to 18 feet apart. The larger growing varieties are usually planted farther apart. Usually about 100 trees to the acre is the proper number, which means the trees must stand in the neighborhood of 18 to 24 feet apart.

Harvesting and Curing.—The citrus fruits are harvested throughout the year. There is considerable difference observed in the picking and the packing of the different citrus fruits. Some fruits are picked while they are green and allowed to cure before they are shipped, while others are picked as soon as they are ripe and shipped at once. All citrus fruits must be handled with care, and precaution taken to see that thorns do not fall into the picking vessel. Such thorns or sharp twigs will scratch the skin of the fruit and damage it.

In picking citrus fruit some precaution should be taken to see that no imperfect specimens are included in the package. The fruit must be separated from the tree by means of a clipper which cuts the stem off close to the fruit. The picked fruit should be placed in baskets or crates. The fruit should be taken to the packing house with the greatest care and permitted to cure before it is fit to pack for shipment. After the fruit has been picked for some time the skin will toughen and the fruit will shrink, and then it can be handled with less danger of being injured. The curing time varies with the different citrus fruits and ranges from several days for the orange to several weeks for the lemon. After the fruit has cured properly it is graded and packed.

The oranges are harvested throughout the year. The Valencia are the summer oranges and they are harvested from June to November, while the Navel or winter oranges are picked from November to May. The seasons for both oranges somewhat overlap.

The oranges should be picked with a great deal of care so that all bruises or cuts on the skin will be avoided. Any abrasion of the skin admits the germs of decay and the fruit is ruined.

The oranges in some cases must be colored or cured by sweating. The sweat-room is an air-tight, fire-proof room built separately from the main packing-house. The heat is provided by kerosene stoves which do not give complete combustion. The hot gases and water vapor fill the sweating room and envelop the fruit. The temperature is controlled by ventilators. In the sweating process the fruit is kept at a temperature of 100° F. The time of curing varies from three to five days or until the oranges are properly colored.

The harvesting and the curing of the lemon differs greatly from that of the orange. The lemons are usually picked from ten to twelve times a year. The heaviest pickings of the lemon come in March and April, while the lightest pickings come in August and in September. This roughly divides the lemons into a fall and a spring crop. The summer crop is usually rushed to the market while the winter crop is held until later in the season.

The method of picking the lemons from the tree is similar to that of the oranges. The chief difference between the two fruits is that the lemons are always picked by the use of a ring. The lemons are harvested while green and therefore a ring is used to determine the size and maturity of the fruit. The picking rings are made of iron wire. The rings vary slightly in size and during the summer a ring 2 1/4 inches in diameter is used while during the winter a larger size, namely, 2 5/6 inches, in diameter is used. The larger ring is used in the winter because the fruit will be kept longer and a greater amount of shrinkage will take place.

After the lemons reach the packing-house they must go through a curing process. The curing is done by subjecting the fruit to a sweating. The sweating of the lemons is for the purpose of quickly changing the green color to a light yellow color.

The lemons should be sweated alternately for the best

results. The air of the sweating-room should be kept saturated with moisture all the time. If the air is allowed to become dry the lemons shrivel quickly. The temperature of the sweating chamber should be held around 90° F.

The winter lemons are usually stored and held for spring trade. Therefore the winter lemons are not treated in the same manner as the summer lemons. Instead of sweating the fruit and hastening the curing the lemons are prevented from sweating. As soon as the fruit is brought to the packing-house it is washed in a very weak solution of copper sulphate. This solution is made by adding 1 pound of copper sulphate to 1000 gallons of water in the morning and 1/2 pound at noon to keep the strength constant. The fruit is washed in this manner for disinfection against the brown rot. The lemons after being properly graded are placed loosely in packing boxes and stacked up on the storage floor. Lemons are often stored in this manner for six or seven months.

FIG. 126.—Showing the method of washing oranges to remove the sooty mould fungus. (Bulletin No. 123, United States Department of Agriculture, Bureau of Plant Industry.)

Washing.—In most of the citrus-growing regions the fruit must be washed before it is shipped. If the fruit is grown on a healthy tree, free from diseases or scale insects, washing is not always necessary. The appearance of the fruit cannot be improved by washing unless it is grown where it is dry and windy and the fruit is covered with dust. If the fruit is affected with sooty mould which causes black spots it must be washed.

The fruit is either washed by hand or by machine. Various machines have been invented for this work. A machine which gives good satisfaction is made with a series of brushes. The brushes are slightly larger than scrubbing brushes and are arranged on a chain belt. The fruit is placed on a chute and rolls into a Vessel containing water. It is then made to circulate in this vessel between the brushes, and in this way is cleaned. There are several other washing machines, but all are constructed on the same general principles.

Grading.—The citrus fruit cannot be packed as it comes from the orchard. It must first be graded. All fruits of one size should be sorted out and placed in a given bin. Most of the grading is done by machinery. Scattered along the belt which carries the fruit when it is graded are several men whose duty it is to take out all of the imperfect or defective fruits. The remainder are carried along the belt until they reach the opening of the proper size, where they fall through and are caught in a bin. This method of grading saves time and labor. The fruit of different sizes is collected in a separate bin and can then be packed in a uniform manner.

The lemons are usually picked with a ring, which makes them approximately one size, and very little if any grading is necessary. The orange requires more grading perhaps than any other citrus fruit.

Packing.—The citrus fruits are packed either by hand or by machinery. By far the greater percentage is packed by hand. Each fruit is placed in a given position in the box, and uniform packing has been developed to a high degree.

The fruit is wrapped in paper. Sometimes a monogram or some other pleasing design is printed on the wrapping paper.

The fruit is then packed in boxes and the number of fruit in each box is determined by the size of each specimen. The number is always the same for a given size.

The oranges which are suitable for packing vary in size from 2 1/8 to 3 1/2 inches. The orange crate measures approximately 12 × 12 × 28 inches. This crate holds 360 specimens of the smallest size and only 80 of the largest size.

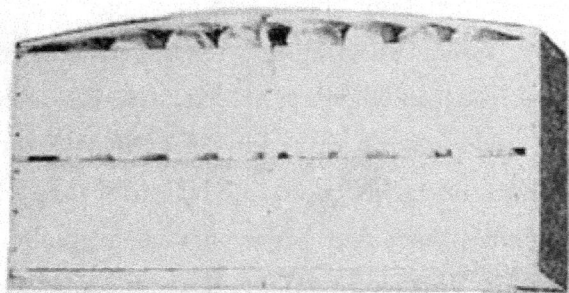

FIG. 127.—The usual package for citrus fruit.

The grape fruit is packed similar to that of the orange. The picking season for this fruit ranges from December until the following August. The fruit is ordinarily stored in boxes for several days, until the skin becomes soft. After the skin has reached the proper stage the fruit is wrapped in paper and packed in boxes the same as oranges.

DISEASES OF THE CITRUS FRUITS

The citrus trees are susceptible to the attack of a number of diseases. The fungus diseases as well as the physiological troubles cause great loss to the citrus grower. In regions where the climate is moist the damage from fungus disease is greater than in regions where the climate is dry. The reverse is true with regard to the physiological troubles.

The disease injury to the citrus tree is found on the root, the stem and the fruit, and in this respect resembles the injuries found on many of our temperate fruits.

Gum Diseases.—The citrus fruits often secrete gum from many parts of the tree, due to a number of causes. It seems to be the direct result of certain forms of diseases, and generally such troubles are classed as gum diseases.

The leaf gumming is very common, especially on the orange. It is more prevalent when the weather is very warm. The gum appears as little drops, usually on the undersides of the leaves. It is reddish brown in color. This trouble is not very serious and should cause no uneasiness.

The brown rot gumming is caused by the brown-rot fungus. It is most common on lemon trees. The greatest exudation of the gum usually occurs on the trunk of the tree close to the bud union. This disease can be largely prevented by avoiding soil conditions which are the most favorable for the growth of the fungus. Do not allow water to stand around the tree or to come in contact with the trunk.

Twig gumming is sometimes found on nursery stock. It is thought to be caused when the trees are copiously watered after they have dried out considerably. The gum is found on the twigs and causes the bark to split. The leaves usually drop and the twigs die.

Rots.—Besides the gum diseases there are several rots which are injurious. The foot rot and the toadstool rot are the most important. In Florida the foot rot is well distributed, but it is comparatively rare in the citrus belt of California. The root rot is the result of a fungus which causes the roots to rot. The affected roots soon become soft and slimy and the disease gradually spreads downward. The sour orange is the least susceptible to the attack of the fungus, and the disease is largely controlled by grafting on the sour orange stock.

Toadstool Rot.—The toadstool rot is the result of a fungus growth. This fungus is native to the root of the oaks, but it has been able to flourish on the citrus trees, and it is causing much damage. The disease usually kills the tree in three or four years, the affected tree dying gradually. During a long, rainy season this fungus produces several clusters of brownish colored toadstools from the roots. It is from these toadstools produced as the fruiting bodies that the fungus take the name. At present there is no satisfactory remedy for this disease.

Brown Rot.—The brown rot of the fruit causes great losses annually. The spores of this disease enter the fruit through the breathing pores, where they germinate and grow in the fruit. In a short time the fruit begins to decay and it soon develops the characteristic brown color. All of the citrus fruits are affected with this rot, but the lemons suffer the least from its attacks. The loss from this disease is the greatest during wet weather. The disease continues to spread rapidly in the packing house and often destroys whole boxes of fruit before it is detected.

The brown rot can easily be controlled if the fruit is washed in water which contains copper sulphate at the rate of 1 1/2 pounds to 1000 gallons of water.

Stem-end Rot.—The stem-end rot affects the stem of the fruit and causes it to drop. The dropping begins with the green fruits and continues through the entire season. The stem-end rot often causes the fruit to decay after it has reached the market. This disease is difficult to control, but the most successful method of control is to keep the tree carefully pruned and to remove and destroy all mummied fruit and dead twigs.

Mould.—The blue and the green mould of the fruit causes great losses in the citrus industry. These moulds are only slightly parasitic on perfect fruits, and the decay is confined principally to those fruits which have been injured in handling. The moulds produce a soft rot and the spores appear as blue or green powder on the surface of the affected fruit. The loss from this disease can be largely prevented by the careful handling of the fruit.

Several other less important diseases are found on the citrus fruit, and the reader is referred to some more extensive treatise of that subject should he desire more information concerning citrus diseases.

INSECTS OF THE CITRUS FRUITS.

The insects which attack the citrus fruits are small in number, but they are very resistant to any control measures. They cause great financial loss annually. The scale insects are the most widely distributed and probably cause the greatest damage.

The amount of the insect damage is largely controlled by the climate. In one region a certain insect pest will predominate and do great damage while in another region a different insect will do the greatest damage. Each citrus region usually has some well-defined insect which may not be serious in any other region.

FIG. 128.—An orange tree partly killed by the red scale.
(After Quayle,
California Agriculture Experiment Station.)

The control and eradication of all citrus insects is based

principally upon sanitation. All weeds should be destroyed. The fence rows should be clean and rubbish which harbors insects should be removed. Where perfect sanitation exists and a systematic and a logical program of fumigation is practised no great amount of damage is caused by the insect pests.

Scale Insects.—The scale insects which arc troublesome, in almost every case, are foreign insects which have been introduced into this country through shipments of nursery stock and by other ways. Some of the scale insects not only damage the plants by sucking out the juices, but they secrete a sweet substance which gives a good medium for the grapth of certain moulds.

FIG. 129.—Citrus trees covered with tents preparatory to fumigating them, taken at night when the operation is carried on. (After Quayle, California Agriculture Experiment Station.)

27

The most practical means of control of the scale insects is by fumigating with hydrocyanic acid gas. Each insect varies in its power to withstand the gas and separate dosage tables have been worked out for each important scale insect. The success of this gas in controlling the scale insect is in its ease of generation and its exceedingly poisonous nature.

The fumigation is done by the use of a tent placed over the tree. The tents are made of the best duck and vary in size from 20 to 36 feet for different sized trees.

The cost of fumigating is about thirty cents for the average-sized tree.

Thrip.—Besides the scale insects several others are injurious. The orange thrip is often troublesome and is found principally in the flowers of the citrus fruit. The presence of the thrip is usually first detected by the distorted and irregular growth of young leaves.

The thrip injures the fruit by producing irregular scars around the stem and at other places over the surface. The damage done to the fruit does not injure the edible qualities, but it reduces the sale and places it in an inferior grade.

The most effective remedy for the thrip recommended by the United States Department of Agriculture is 2 1/4 quarts of commercial lime sulphur at 22° Baumé plus 3 1/2 fluid-ounces of a 40 per cent. Black Leaf extract to 30 gallons of water. This material should be sprayed on the trees with a force of 175 or 200 pounds pressure.

Red Spider.—There are two species of red spider injurious to the citrus fruit. They are found throughout the citrus regions both in Florida and in California.

The red spider is a small red insect which often becomes so abundant on a leaf as to give a reddish color to it. The best remedy for the red spider is sulphur. It is used in either the dry form or in the form of lime sulphur solution. When it is used dry the sulphur is dusted on the plant, usually when the foliage is a little damp. When lime sulphur is used the commercial product is diluted 1 gallon to 35 gallons of water. The lime sulphur is becoming more popular and its cost is much less than fumigation.

Whiteflies.—The whiteflies are serious pests to the citrus industry especially in Florida. In fact they are perhaps among the most destructive insect pests in that region. There are eight different species which may be found on citrus fruit in Florida. The common citrus whitefly, the woolly whitefly and the cloudy-winged whitefly are the three most important species, while the mulberry whitefly, guava whitefly, sweet potato whitefly, bay whitefly and the flocculent whitefly are found in this region but are of minor importance.

The whiteflies are small insects, usually white in color, although some species have blotches on their wings which give them a cloudy appearance, while a few others have markings on their wings which produce a slight variation and give a different appearance.

The life histories of all of the whiteflies are very similar. These insects have complete metamorphoses, that is, they pass through four stages in their development.

The egg is small and nearly oval in outline. It is barely visible to the naked eye. The eggs are laid on the under side of the leaves of the plant and hatch in a few days into, pale yellow larvæ. The larvæ crawl about for a few hours, usually on the underside of the leaves, and soon insert their beaks into the tissue, where they suck out the sap. They never move again, but continue to feed and grow rapidly, going through several molts until they reach the fourth larval stage, which differs from the larva of the other stages in being thicker, taking less food, and the organs of the adult fly begin to form. The insect passes through this last stage and emerges as an adult fly. The life as an adult is very short, in most cases only a few days. The females lay about a hundred eggs and then die.

The whitefly can be held in check by spraying. Spraying with a combination of soap and oils will serve to control these insects. The original spray recommended by the Florida Agricultural Experiment Station is as follows:

Whale oil soap	8 pounds
Paraffin oil	2 gallons
Water	1 gallon

More recent experiments have demonstrated that by substituting other soaps which are cheaper, just as good results are obtained. The modified formula then is to substitute two pounds of ordinary soap for the whale oil soap and then heat the material to the boiling-point and emulsify it by agitating it vigorously. This stock solution should be diluted before it is used.

The number of sprayings that are necessary to control the whiteflies depends upon the weather and the location. The first spray should be given late in April or early in May. The best time seems to be when the bulk of the spring brood of adults have disappeared. Several other sprayings are usually necessary and are determined by the location. The exact time to spray should be determined by the flies and should be about two weeks after a great number of the adults have disappeared. Usually spraying must be done in late August and early September. This spray is considered the most important of the year.

The whiteflies are also attacked by certain fungi. These fungi have proved very effective in holding the insects in check. When the summer rains begin, which is usually in June or July, an attempt should be made to introduce these fungi in the citrus grove which will operate as natural enemies for the control of the whitefly.

Control of Insects.—The control measures for the citrus insects are different from the control measures for most other insects. This is made necessary because the trees have their leaves the entire season, and in the case of the scale insects, the insecticide effective in killing the scale would also destroy the foliage. In Florida where the whiteflies are very destructive spraying with miscible oils has proved very effective in controlling these insects. However, owing to the different classes of insects as well as the fact that the trees are in foliage at the time control measures have to be practised, has led to another method which is largely practised in certain citrus regions, namely that of fumigation.

About 1886 California first seriously considered fumigation as a means of destroying injurious insects. As time passed the methods of fumigation were greatly improved but the fundamental principles remain the same.

Fumigation is practised by covering the tree with a tent made of heavy duck. Under the tent the fumigating material is placed. Hydrocyanic acid gas is the material commonly used, and is made by depositing sodium or potassium cyanide in an earthen jar and covering it with sulphuric acid. The jars should be at least 2 gallons in capacity to prevent the acid from foaming up and spilling out. The amount of material which is used depends upon the size of the tree and the insect which is doing the damage. There are dosage tables worked out by the United States Department of Agriculture and the various state experiment stations, and these should be consulted for a more detailed study of fumigation.

The fumigation gives better results and produces less injury to the foliage if it is done at night instead of in the daytime.

The season of the year at which fumigating is done depends upon the life history of the insects and the condition of the tree. The fumigation should be carried on when the insects are in the most tender stage and can be easily killed, and the time will vary slightly for each insect. However, from August to January seems to be the time which gives the most satisfactory results.

The dosages as well as the length of time vary with the different scale insects. This phase can be learned in some more complete treatise on the subject.

REVIEW QUESTIONS.

1. Name the fruits included in the citrus class.

2. Why is the citrus industry confined to certain special districts?

3. How does the grape fruit differ from the orange?

4. Discuss the propagation of the citrus fruit. What form of budding is used?

5. What is the best type of soil for the citrus fruit? How does the subsoil regulate the value of the top soil?

6. Why is good cultivation essential?

7. How should a citrus orchard be planted?

8. Discuss the curing of citrus fruits. How does the curing of the orange differ from that of the lemon?

9. What method of picking is used for the lemon? Why?

10. How does the picking of the orange differ from that of the lemon?

11. What is the value of washing citrus fruits? In what solution is the washing done?

12. Discuss the grading of citrus fruits.

13. Discuss the packing of the lemon, grape fruit and orange.

14. Discuss the principal diseases and give the methods of control.

15. What class of insects is the most injurious to the citrus fruit? Why?

16. Discuss the fumigation of a citrus orchard.

17. Why is fumigation used instead of spraying?

18. When is the best time to fumigate? Why?

19. What is the most destructive citrus insect in Florida?

20. Give the life history of the whitefly.

21. What method of control has proved the most effective with the whiteflies.

22. Discuss the various ways of controlling the different

diseases on the citrus fruit.

www.ingramcontent.com/pod-product-compliance
Lightning Source LLC
Chambersburg PA
CBHW032021190326
41520CB00007B/573